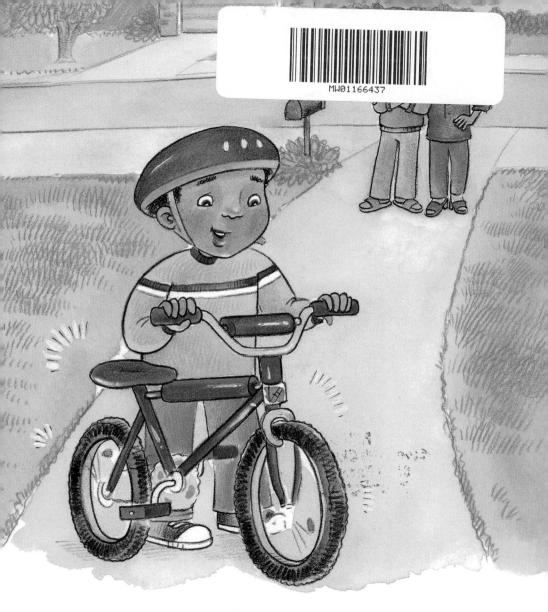

Gabe likes to ride.

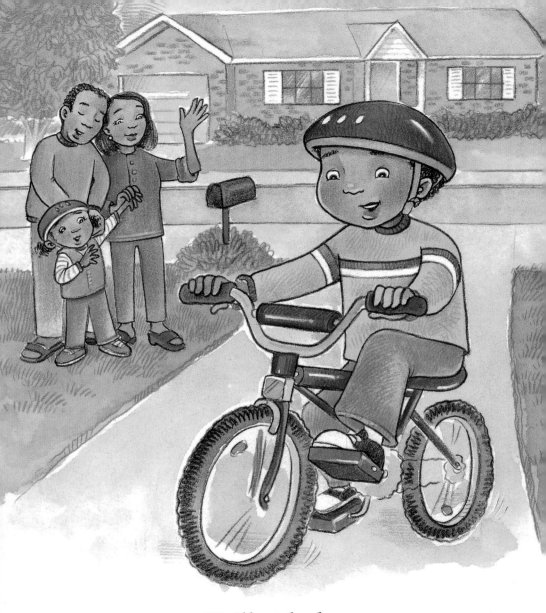

He'll ride here.

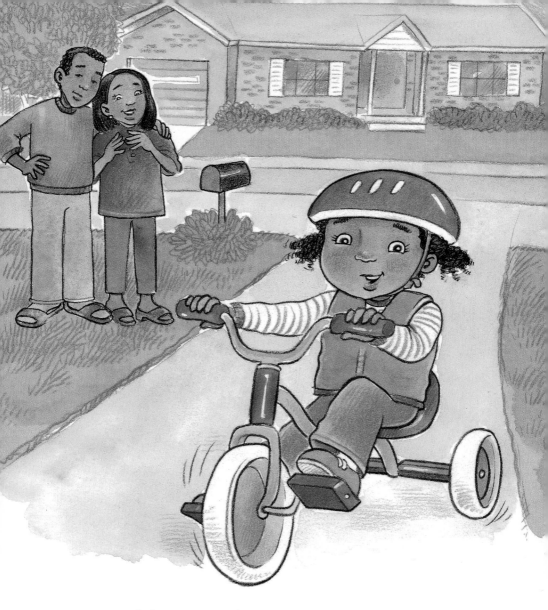

"I like a trike," said Kate.

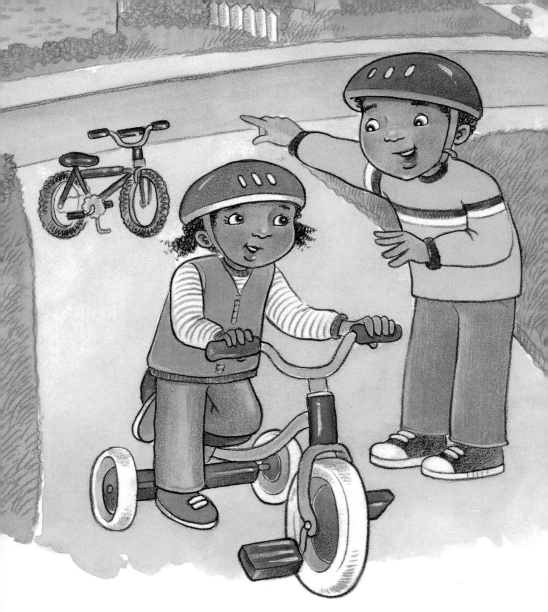
"We'll ride to the store," said Gabe.

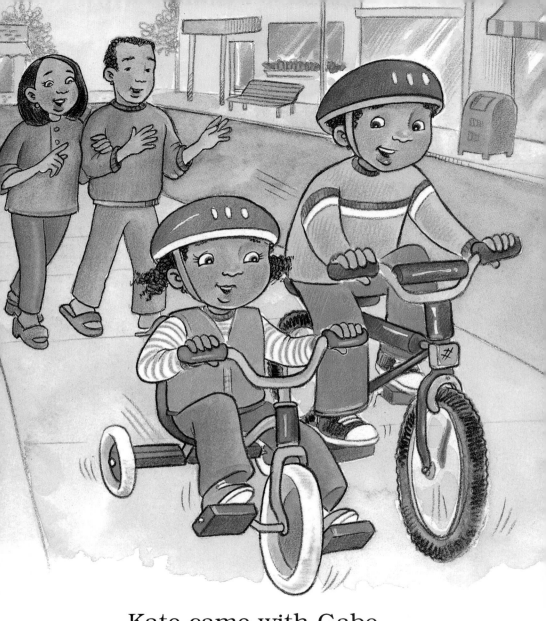

Kate came with Gabe.

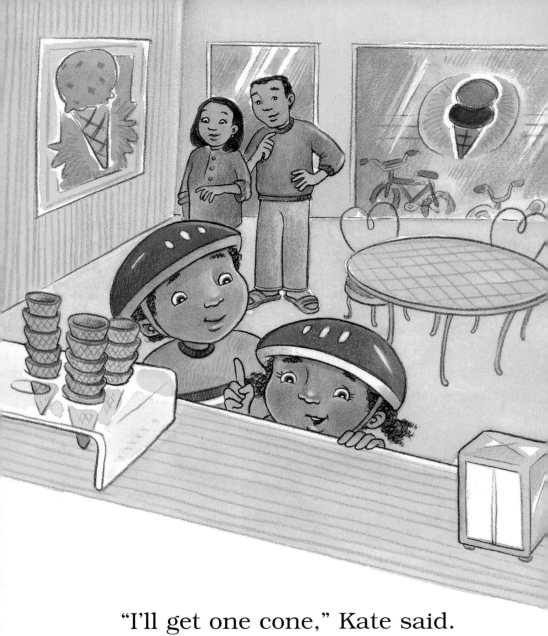

"I'll get one cone," Kate said.

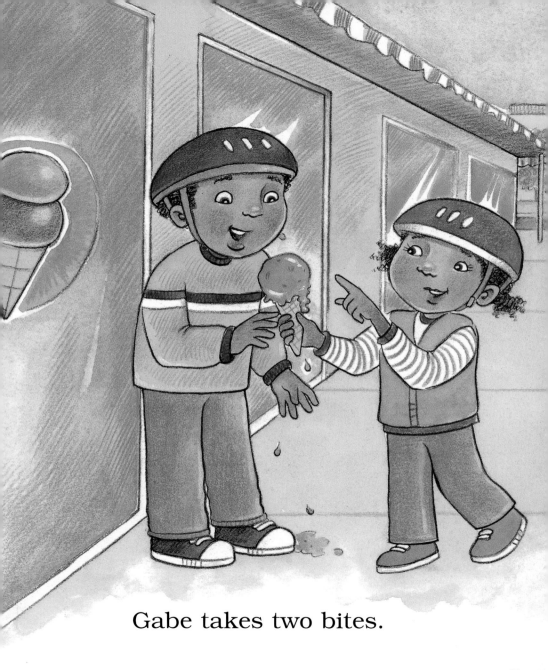
Gabe takes two bites.

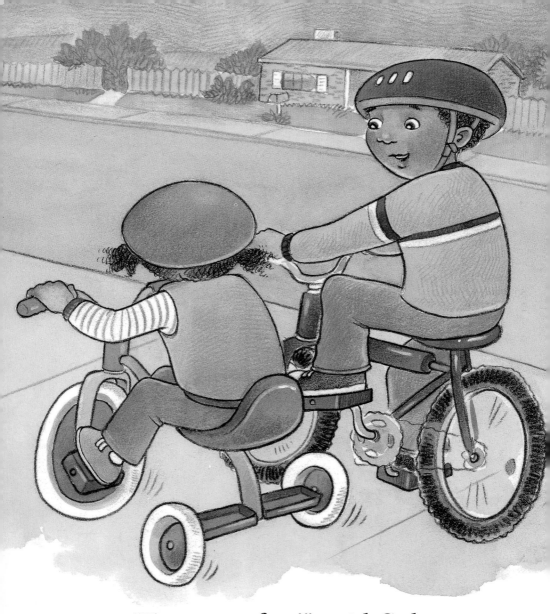

"Trips are fun!" said Gabe.